OSC IB STUDY & REVISION GUIDES
FOR THE INTERNATIONAL BACCALAUREATE DIPLOMA PROGRAMME

osc

Chemistry
Internal Assessment
SL & HL

2nd Edition

NEW FOR 2017

Dave Allen

OSC IB Study & Revision Guides
Published by OSC Publishing,
Belsyre Court, 57 Woodstock Road,
Oxford OX2 6HJ, UK

T : +44 (0) 1865 512802
F : +44 (0) 1865 512335
E : osc@osc-ib.com
W: osc-ib.com

Chemistry
Internal Assessment
SL & HL
2nd Edition
© 2017 Dave Allen
9781910689134
134.01

No part of this publication may be reproduced, stored in a retrieval system, or transmitted in any form or by any means, without the prior permission of the publishers.

PHOTOCOPYING ANY PAGES FROM THIS PUBLICATION IS PROHIBITED.

Dave Allen has asserted his right under the Copyright, Design and Patents Act 1988 to be identified as the author of this work.

The material in this book has been developed independently of the International Baccalaureate Organisation. OSC IB Study & Revision Guides are available in most major IB subject areas. Full details of all our current titles, prices and sample pages as well as future releases are available on our website.

Printed and bound by CPI Group (UK) Ltd, Croydon CR0 4YY
www.cpibooks.co.uk

Preface

"We think there is colour, we think there is sweet, we think there is bitter, but in reality there are atoms and a void."

Democritus, C. 460–C. 370 BC

Welcome to the student's guide to the Chemistry Internal Assessment (IA).

The aim of the guide is to provide advice with regards to:

- understanding the various facets involved in carrying out and writing up a quality lab report; and
- successfully understanding, applying, and implementing the IB's IA assessment criteria.

There is also a short section which contains some possible ideas for IAs as well as a list of dos and don'ts and a list of the mandatory practicals that you will need to carry out at some point during the course.

I should stress that the book is only a guide. It is not a definitive set of instructions or a checklist walking you through a list of prescribed IAs. Your IA will still need you to do some work in terms of planning and implementing the practical. You will need to come up with a viable research question, plan and carry out the investigation, and finally write it up.

Chemistry is a wonderfully interesting subject and it explains to us how the universe works. Studying Chemistry does involve some theoretical work but the real thrill and enjoyment of studying it is that it allows you carry out practical work. This is when you carry out experiments where you will actually be manipulating atoms, molecules or ions.

The IB course dictates that you will carry out 40 hours lab work for SL and 60 hours for HL and much of this will be teacher directed. However, for 10 hours you will have the opportunity (through your IA) to decide on the direction that this lab work takes. It is a sublime opportunity to show your creative side in terms of carrying out and writing up your lab work and can be a very rewarding experience—so when the opportunity presents itself, grab it and make the very most of it!

I would like to thank Dr Caroline Evans, Head of Chemistry at Wellington College, Berkshire for taking the time to proofread this guide and offer me some very useful and constructive advice to make it easier to follow. I would also like to thank the many, many teachers who have attended workshops I have run for sharing their experiences of the IA with me and giving me the drive and inspiration to put this guide together. Finally, a big thank you to Mr Cameron Hunter, Vice Principal at UWCSEA East, for firstly being a great friend and secondly an inspiring colleague who patiently helped me understand the ins and outs of the IA process when it was all still very new to me.

"Chemistry is an experimental science: its conclusions are drawn from data, and its principles supported by evidence from facts."

Michael Faraday, 1791–1867

Contents

Introduction 7
   IB Requirements 7
   What Is the IA? 8
   List of Examples of Practical Work Used Throughout This Book 8

1. Preparing for the IA 9
   1.1 The Scientific Method 9
   1.2 The Research Question 10
   1.3 Variables 12
      1.3.1 Independent variable 12
      1.3.2 Dependent variable 12
      1.3.3 Control variables 13
   1.4 The Method 13
   1.5 Collecting Data 14
      1.5.1 Quantitative data 14
      1.5.2 Qualitative data 14
   1.6 Uncertainties 14
   1.7 Percentage Error 16
   1.8 Significant Figures 18
   1.9 Conclusion 18
   1.10 Literature Value 18
   1.11 Total Percentage Error 19
   1.12 Systematic and Random Errors 19
      1.12.1 Random errors 19
      1.12.2 Systematic errors 19
      1.12.3 Discussion of random and systematic errors 19
   1.13 Evaluation (of the Method) 20
   1.14 Improvements (to the Method) 22
      1.14.1 A note on evaluations and improvements 22
   1.15 Safety and the Environment 23

2. The Assessed Criteria 25
   2.1 Introduction 25
   2.2 How to Use the Assessment Criteria 25
   2.3 Personal Engagement (PE) 27
   2.4 Exploration (EX) 28
   2.5 Analysis (AN) 29
   2.6 Evaluation (EV) 30
   2.7 Communication (CM) 31

3. What Makes a Good IA? 33
3.1 Different Types of IAs 34
3.2 50 Ideas for IAs 34
3.3 10 Dos and Don'ts for Writing a Successful IA 36
Appendix 1: List of Mandatory Practical Work as Specified by the IB 37
Appendix 2: Table of Uncertainty Values for
Lab Equipment Without a Scale 38

INTRODUCTION

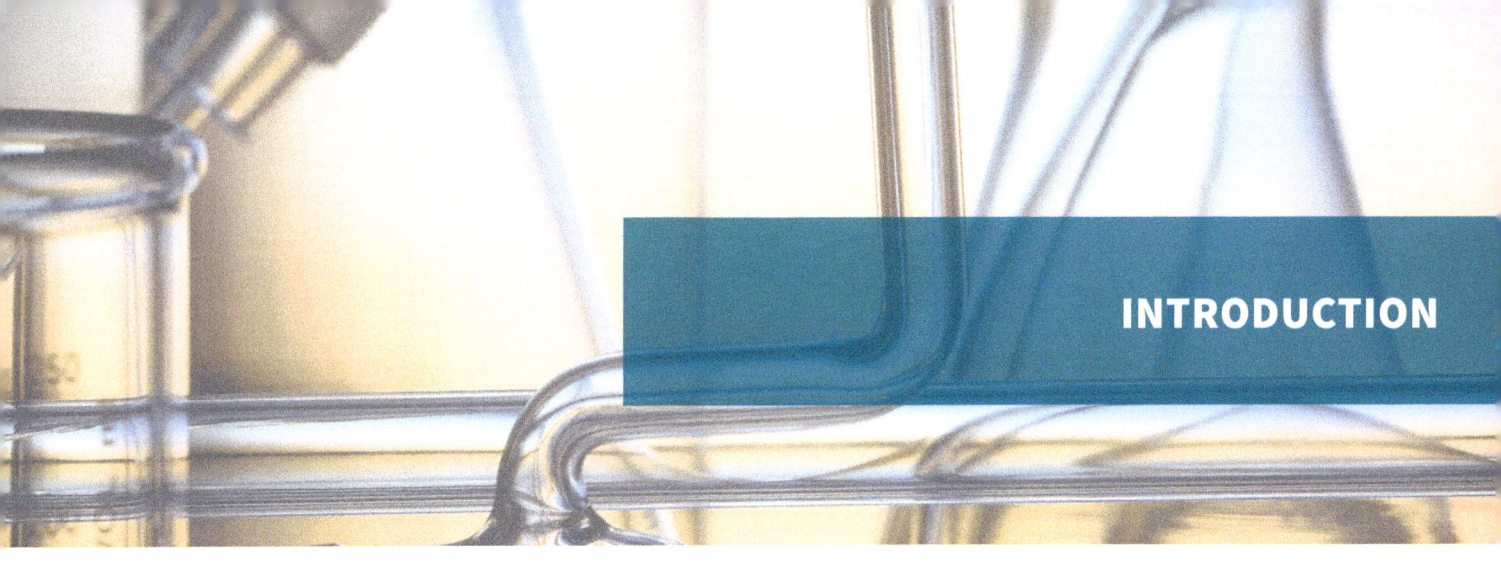

Chemistry is considered to be the central science. It involves the study of, and the interaction of, atoms and provides the link between aspects of the physical world (what atoms are made of) and the biological world (how atoms interact). Without chemistry there would be nothing.

It is also a practical science, and advances in chemical knowledge are only made through practical work.

The art of planning experiments, collecting data, analysing results, and evaluating the procedure are central to the progression of chemical knowledge and it is for this reason that practical work plays a central part of the IB Chemistry course. It goes without saying that exemplary practical skills are a prerequisite for success in chemistry.

Chemistry practical work is responsible for changing the world and this has occurred many times and on many levels. Practical work was behind the extraction of metals from ores many millennia ago (arguably, the earliest example of practical chemistry), to Boyle who discovered the gas laws, and Lavoisier who discovered oxygen, to the industrial production of ammonia via the Haber-Bosch process, a process that has led to the global mass production of fertilisers.

> **Key Point**
>
> By 'practical work', the author means lab work, labs, experiments or experimental lessons / classes.

IB Requirements

There are a number of formal requirements with regards to practical work at diploma level.

To be awarded chemistry points you must carry out 40 hours of practical work at SL or 60 hours at HL over the duration of the IB course.

Out of this required number of hours, 10 hours is set-aside for the Group 4 Project (G4P) and 10 hours for the Internal Assessment (IA). The remainder of the time allocation can be filled by practical work of your teacher's choice but it does need to include some mandatory practical work that will help you to develop good lab skills (for example, determining the molar mass of gas). A more detailed list of the mandatory practical work can be found in Appendix 1.

Whilst not an essential part of the Chemistry course, you may also choose to write an Extended Essay (EE) in chemistry. The EE is an essential part of the Diploma Programme and you will need to write an EE in order to be awarded the diploma. If you do choose to write your EE in chemistry it needs to be very different to your IA or you are at risk of putting your whole diploma in jeopardy. The IA and EE must be very different from each other.

> **Did You Know?**
>
> There are also five different ICT requirements that you should use at least once during the course in conjunction with practical work. These are data logging, data bases, the use of spreadsheets, graph plotting software and simulations.

Introduction

What Is the IA?

The IA is a key part of the Chemistry course and allows you to plan and carry out a miniature research project of your own choice. Naturally, there will be constraints on this, dictated by the availability of equipment, chemicals and apparatus that you require.

It is also an investigation which means that you will need to investigate how a physical quantity or property is affected by a change in conditions.

Your IA should take 10 hours in total and the way the IA is assessed is the same whether you are studying SL or HL. The 10 hours will include planning time, investigation time and write up time.

The IA makes up 20% of your final points score for chemistry.

There are five criteria that are used to assess the IA (personal engagement, exploration, analysis, evaluation and communication). These will be considered in more detail in Chapter 3.

Initially, your teacher will assess your work. The teacher's marks will then be submitted to the IB and a sample of work will be requested for moderation. A moderator will then look at the work in the sample and decide if your teachers grading of the work has been too generous, too harsh or at the correct level. If needed, the IB will use the results of the moderation to adjust the marks of all the work in a school.

> **Author's Tip**
>
> Interestingly, 20% of your final chemistry point score does not equate to 20% of the course teaching time. The IA accounts for only 5.6% of your SL teaching time and 4.1% of your HL teaching time—so this makes it even more important that you give it the time and effort that it is due.

List of Examples of Practical Work Used Throughout This Book

Example	Brief title	The example is used to illustrate…	Page(s)
1	Calculation of activation energy	The research question	11
2	Investigating the rate of evaporation	An introduction to variables	12
3	Investigating electrical conductivity	The independent variable	12
4	Investigating how surface area affects rate of reaction	The independent variable	12
5	Investigating rusting	The dependent variable	12
6	Investigating an enthalpy change	Percentage error	17
7	Determining the enthalpy of neutralisation	Discussion of random and systematic errors	19
8	Preparation of lead (II) iodide	Evaluation of the method	20

1. PREPARING FOR THE IA

TOPICS:
The scientific method
The research question
Variables, The method
Collecting data, Uncertainties
Significant figures, Conclusion
Systematic and random errors
Evaluation, safety and the environment

The IA is not something that you can just expect to turn up on the day and do. You will need to think very carefully about what you choose to investigate and how you intend to carry out your investigation.

It should be considered to be like a computer 'background' programme. It is something that is running all of the time and never closes down. Your IA should also be like this and in your chemistry classes you never know when something may come up that you can use as a starting point for your IA when the time comes.

There may be a practical that you carry out and you start wondering how you could extend this or develop the ideas it is covering further, or there may be a class discussion that stimulates you. These are the 'light bulb' moments that you need to keep and hold onto you—and not forget about!

The following terms are all important facets of the IA and key to obtaining a good IA grade. Please take time to read and familiarise yourself with them prior to starting your IA.

1.1 The Scientific Method

The scientific method is the approach we use to ask and answer scientific questions. It is the method used to discover new ideas or to develop and challenge models.

The scientific method has process been refined over many years but the cycle of experiment → observation → hypothesis → prediction is essentially the same now as it was millennia ago.

> **Author's Tip**
>
> If you make notes by hand it is recommended that you dedicate the first page of your file or book to 'Ideas for My IA'. If you use digital media to make notes (for example, Google Docs or OneNote), create a page and title it the same, i.e., 'Ideas for My IA'. Make sure you jot down those light bulb moments accordingly.

> **Critical Thinking**
>
> It also fits in with Theory of Knowledge where the scientific method should form part of the study of Natural Sciences which itself is one of the Areas of Knowledge.

It can be visualised in the following diagram:

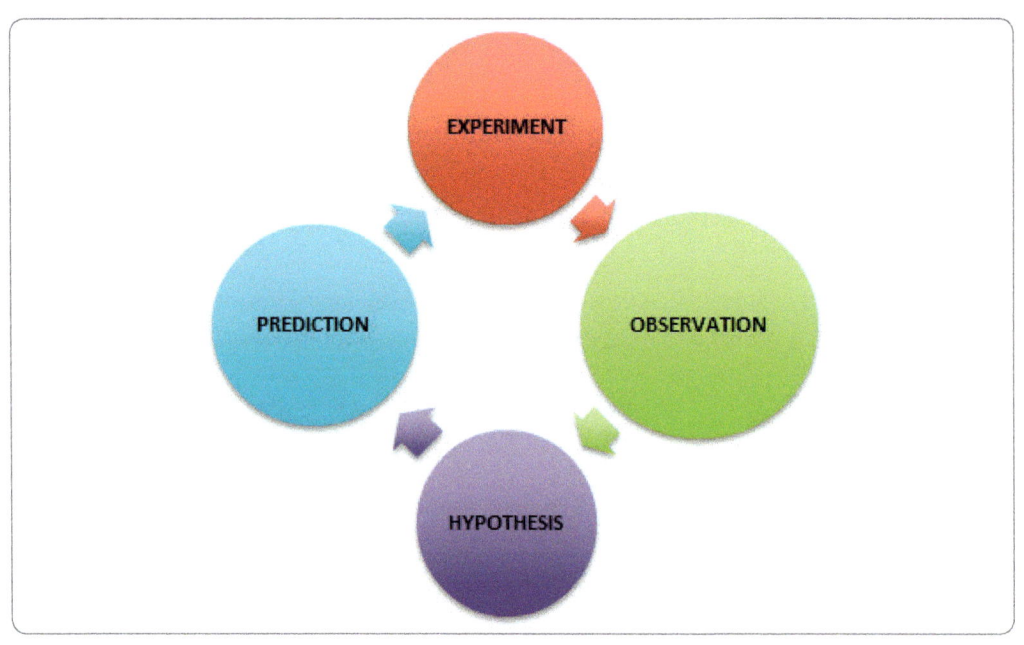

Figure 1.1: *The Scientific Method*

A hypothesis can be thought of as an 'educated guess' that uses prior knowledge and understandings to explain an observation.

On the other hand, a prediction is an 'educated guess' based on what might happen to the observation if another factor is applied to it.

This is best illustrated through an example:

> **Example:**
> **Observation** = spilt water will evaporate faster on hot days as opposed to cold days.
> **Hypothesis** = the liquid water molecules are gaining energy from the environment and using this to turn into water vapour.
> **Prediction** = on hotter days the water molecules gain more energy so the evaporation will occur faster.
> **Experiment** = test the speed of evaporation of water at different temperatures.

There are no entry and exit points to the scientific method and the cycle is continuous—it never ends. It allows ideas and understandings to be continually refined to generate a deeper understanding of things.

That said, for your IA it would be advisable to start with the observation as this will allow you to come up with some ideas based on your current understanding of chemistry. You will then be able to make a prediction and to see how this changes the hypothesis. It is also unlikely that you will go around the cycle more than once.

1.2 The Research Question

Much like the EE, the research question is the backbone of the IA. A good, well-developed and focused research question is more likely to gain a better mark than a poorly focused research question. You should be prepared to invest a good amount of time in the research question, and it is unlikely your final research question will be identical to your original one.

As a rule of thumb, the research question should generally be in the format of 'how does changing *x* affect *y*?' where *x* = the independent variable and *y* = the dependent variable (see the section on pages 12-13 concerning variables).

It will allow your IA to be well planned, allow for good collection of data that can be analysed and processed and allow a solid conclusion and evaluation to be made.

It is very hard to immediately come up with a focused research question, and it is far better to slowly develop one as you progress through the IB course.

This process is best explained through Example 1.

> **Example 1:**
> Rosie and Freya are students studying in Australia and Canada. Both are interested in rates of reaction and they have both carried out some practical work looking at the effect of a catalyst on the rate of reaction.
> Both students know from their Maxwell–Boltzmann diagrams that if they use a catalyst the rate of a reaction increases due to the particles having an alternative route in which to react of lower activation energy.

Freya wonders if she can use different catalysts to look at the decomposition of hydrogen peroxide solution.

Her research question is:

> *'Do different catalysts change the breakdown of hydrogen peroxide?'*

Rosie, on the other hand, wonders if she would be able to use the idea of activation energy to determine the activation energy of a reaction for a catalysed and un-catalysed reaction.

A quick piece of research tells her that she could investigate the reaction between potassium (VII) manganate and ethanedioic acid with or without the use of the manganate (II) catalyst.

Her focused research question becomes:

> *'What is the percentage decrease in activation energy of the catalysed and uncatalysed reaction between 0.5 mol dm^{-3} Ethanedioic acid and 0.02 mol dm^{-3} Potassium Manganate (VII) using 0.01 mol dm^{-3} Mn^{2+}(aq) as the catalyst when compared to the uncatalysed reaction?'*

Now, both these research questions satisfy the rule above (how does changing *x* affect *y*?) but it is clear to see who has written the most focused research question.

Freya's question lacks detail and direction. It poses unanswered questions such as what does the hydrogen peroxide get broken down into? What catalysts are used? What concentration is the hydrogen peroxide? Are the catalysts homogeneous or heterogeneous? What concentration or masses of catalysts are used? What data will be processed? And so on. After reading the research question we are still not clear what the experiment will measure.

Rosie's question on the other hand gives us a real snap shot of what the experiment is about. We can see that she is going to calculate the activation energy, the reagents being used, the type of catalyst and so on.

If you think this section is long you are correct—it reflects the importance of the research question.

Author's Tip

A good research question will help you to score well in 75% of the assessed criteria.

Author's Tip

How good is your research question? As a rule of thumb, you should be able to give it to a fellow student, preferably a chemist, and they should be able to read it and understand exactly what you are intending to do. If they have questions about the research question, this tells you that it is not focused enough.

Author's Tip

Practise writing research questions for practical work you have already carried out with your teacher. It is unlikely that there will be a research question for this work and it gives you a good opportunity to practise writing focused research questions.

Author's Tip

The research question should be a question. It should end with a question mark.

1.3 Variables

Variables are conditions or factors that could be changed in an experiment. Consider the next example.

> **Example 2:**
> In an experiment where you were going to investigate the rate of evaporation of liquids, variables would be factors such as type of liquid, amount of liquid, surface area of liquid, temperature of liquid and so on.

When designing an experiment it is important to understand that there are three different types of variables: the independent variable, the dependent variable and controlled variables.

1.3.1 Independent variable

Your IA should investigate one variable only. Note that the subheading is singular, not plural. Don't over complicate the investigation by trying to investigate too many different things. Keep it simple but elegant.

> **Author's Tip**
> The variable that you choose to measure is called the Independent Variable. Try to remember 'I' choose it and 'I' for independent.

> **Example 3:**
> You have chosen to investigate the electrical conductivity of a solution when different salts are dissolved into it, so in this experiment the independent variable would be the type of salts that you add.

You should always ask yourself if you can put a number to the independent variable—this will help you ensure that the research question is well focused. In other words, the independent variable needs some quantitative values.

> **Example 4:**
> If you were investigating how changing the surface area changes the rate of a reaction, you need to have some actual surface area values ($50cm^2$, $100cm^2$, $150cm^2$ and so on) as opposed to values such as 'medium surface area', 'small surface area', 'powder' and so on.

As a rule of thumb, you would aim to change the independent variable five times. Using the above example, you would need to use five different types of salts—but please remember, this is only a rule of thumb and is not a formal IA requirement.

1.3.2 Dependent variable

The dependent variable is the variable that you are going to record or measure. Using Example 2 above, the dependent variable would be the electrical conductivity of the solution once the salt has been dissolved in it. It should be noted that the dependent variable may not change, for example, if you were calculating the activation energy of a reaction.

As with the independent variable, you should always ask yourself if you can put a number to the dependent variable. This means that the dependent variable should be a quantitative measurement. This will help you ensure that the research question is well focused as illustrated using Example 5.

> **Example 5:**
> If you were investigating rusting (see Figure 1.2), it would be no good to look at some samples of iron under different conditions and decide nail number 2 has rusted the most and nail number 9 the least. This is a purely qualitative observation.

1. Preparing for the IA

You would need to have the mass of the nails before and after to be able to show which nail has acquired the most rust. This would allow you to have some numerical data.

Figure 1.2: *Nails at end of experiment*
Source: *Josh Madison, https://joshmadison.com/2003/12/14/will-coke-dissolve-a-nail-experiment/ [accessed 15 Feb 2017] (CC BY 4.0).*

1.3.3 Control variables

These are variables that you deliberately keep the same to ensure that the dependent variable is only affected by one factor: the independent variable. In Example 2 (page12), it would not be a very good experiment if you changed the type of salt but also used different liquids. This may seem an extreme example, but it is a valid one that illustrates the point well, as you would not be sure if it were the type of salt, type of liquid, or both influencing the conductivity of the solution.

1.4 The Method

Think of the method as being a chemical recipe. Ideally, it should be a set of numbered instructions that any IB chemist should be able to follow in order to collect to achieve the same results as you. It is the same as following a recipe for making a cake—you are able to make the same thing as the original chef.

For example, 'pour out some acid' is not a great instruction. 'Measure out 25.0 cm^3 of 1.0 mol dm^{-3} hydrochloric acid using a 25cm^3 pipette' is much better.

The method should allow for relevant raw data to be collected. Using the example of determining the electrical conductivity of a solution when different salts are dissolved into it (see section 1.3.1 Independent variable on the previous page), if an insoluble salt were selected the method would not allow relevant raw data to be collected.

Your method should also include instructions on how to control the variables. It is imperative that you explain how you will change the independent variable, measure the dependent variable and fix the control variables.

Ideally, the method will also have an apparatus list and list of chemicals required for the experiment. It may also include photographs or diagrams that will aid interpretation of the written instructions. All illustrations should be fully labelled. If you are using someone else's diagram, ensure that it is fully referenced.

> **Key Point**
>
> Your instructions should be clear, concise and precise, making use of the correct chemical terminology.

1.5 Collecting Data

Once you have planned your method, you should also plan how you will record your raw data.

The agreed way of doing this is to use a table with the first column containing the independent variable as a heading and the second column the dependent variable as a heading. Don't forget to include units—a classic mistake.

If time permits, it is advisable to repeat your experiment and so there may be more than just two columns of raw data. Repeating your experiment is good practice and it will help you to decide if your results are precise and accurate.

You may also decide to include some processed data in the table, for example, calculation of an average.

During most experimental work, there should be two categories of data that you collect, quantitative and qualitative data. This may not always be the case but for your IA is advisable to try and collect both types of data. Your table of data should contain both types of data.

1.5.1 Quantitative data

Quantitative data contains numbers. For Example 3 cited in the variables section (investigating different electrical conductivities of a solution when different salts are dissolved into it) the quantitative data will be the voltage generated (e.g., 2.85V). There will usually be some number crunching (calculations) involved at a later stage of the lab report using this type of data.

1.5.2 Qualitative data

This is data that can be seen, smelt or felt. It is descriptive and subjective. It can be much harder to analyse. Using the example above, qualitative data may be an observation that some salt sank and did not all dissolve. Qualitative data does not have to form part of the table. It is acceptable for it to be a subheading under the table of quantitative data.

1.6 Uncertainties

This will be dealt with in more detail by your teacher as the subject material is covered in Topic 11 of the course. However, as errors and uncertainties form part of good practice in all practical work, and are also important in your IA, it would be appropriate to include a section on them.

Whenever you make a measurement in an experiment there will always be a degree of uncertainty about it. It is universally agreed that it is not possible, for example, to measure out exactly 10cm^3 of liquid or record an exact temperature such as 42.5°C.

The quality of the measurement depends very much on the precision of the apparatus used. If you were using a 100cm^3 beaker or a 50cm^3 measuring cylinder (see next page) to measure 25cm^3 of water, you would feel less confident about the 'true' volume you were measuring with the beaker than the measuring cylinder. If you used the beaker, you may even think to yourself something along the lines of, 'well that's 25cm^3 give or take a few cm^3'. However, you would feel much more confident that the volume measured with the measuring cylinder was accurate.

> **Did You Know?**
>
> Don't underestimate the importance of qualitative data. Big scientific breakthroughs have been made through qualitative observations, for example, the discovery of penicillin by Alexander Fleming was initially made through a qualitative observation.

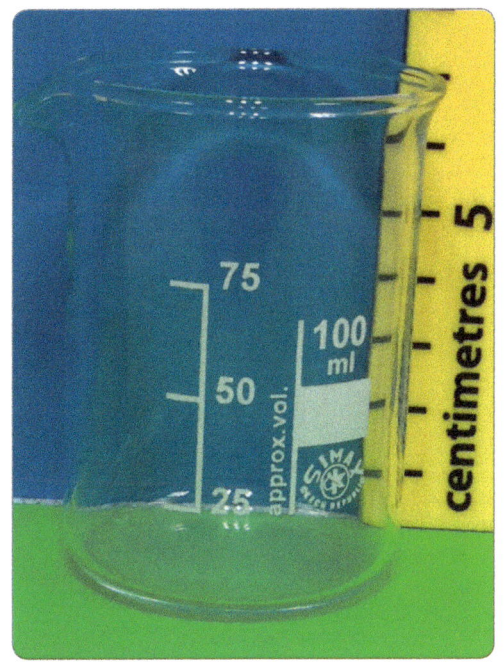

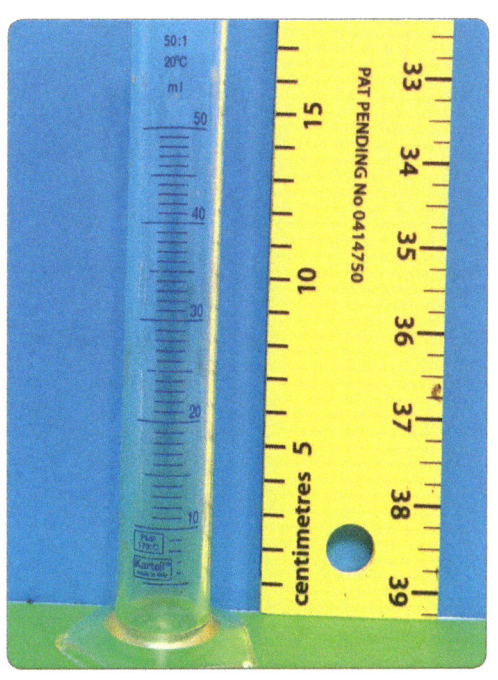

100cm³ Beaker   *50 cm³ Measuring Cylinder*

That said, it would be naive to think that the measuring cylinder will give you exactly 25cm³.

The uncertainty can be thought of as being a level of confidence. The accepted way of determining most uncertainties is to quote half the smallest unit you can measure. In the measuring cylinder the smallest measurable unit is 1cm³ so the uncertainty would be agreed to be ± 0.5cm³ and so the measured volume would be 25.0cm³ ± 0.5cm³. Note that we quote the volume to the same number of decimal places as the uncertainty (i.e., 1 dp) so it becomes 25.0cm³ and not 25cm³.

Sometimes, due to the apparatus you have, it is acceptable to estimate the uncertainty. For example, if the only piece of apparatus you had to measure the volume of liquid was a beaker, it would be acceptable to estimate the uncertainty, as the scales are far less precise and an estimation is probably more precise. In this example, you may decide that it would be acceptable to quote, 25cm³ ± 3cm³. In this example, you would be advised to make a note of the fact that you have decided to estimate the uncertainty and explain why you did this.

A more common example of estimating uncertainties comes from measurements that involve time. If you use a stop watch on your phone, it is likely to give you a value to the nearest 1/100 of a second, but are you able to stop the clock this quickly? In this example, it would be far better to estimate the reaction time, possibly by seeing how quickly you can start and stop the clock.

Would an estimated uncertainty of ± 0.15 sec be more realistic than using the smallest value (i.e., ± 0.01 sec)?

It is not possible to determine uncertainties for some pieces of apparatus when there is only one measurement, for example, a pipette or a volumetric flask.

With this type of apparatus, you will need to either consult your teacher, a table of data or the apparatus itself (the value may be printed on the apparatus).

Some of these pieces of apparatus are 'A' grade or 'B' grade, depending on the

1. Preparing for the IA

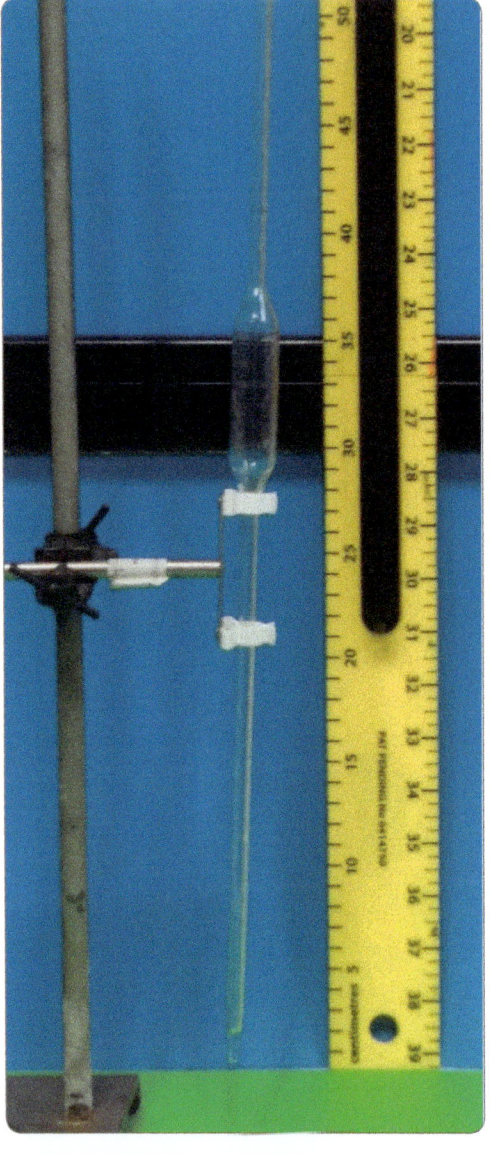

quality. 'A' equipment is more common in universities or research laboratories than schools. 'B' grade apparatus is much more likely to be found in schools.

A 'B' grade pipette has an uncertainty of $25.00cm^3 \pm 0.05cm^3$ and a 'B' grade volumetric flask of $100.0cm^3 \pm 0.1cm^3$. Some more typical uncertainty values are found in Appendix 3.

When a piece of equipment has a scale, we can consider it to be an 'analogue' piece of equipment.

If the piece of apparatus is digital such as a balance (or thermometer), the accepted uncertainty is the smallest value. So the mass of the crucible in the photo below would be $54.678g \pm 0.001g$.

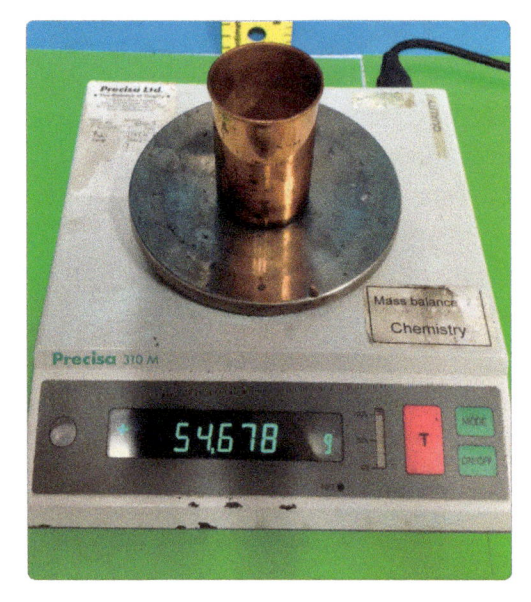

1.7 Percentage Error

The real benefit of determining the error of a piece of equipment is that it allows you to determine percentage error specific to a measurement and, ultimately, the overall percentage error.

To calculate the percentage error for a specific measurement use this calculation:

(Uncertainty value of the apparatus / quantity measured) x 100

If, $25.0cm^3$ of water had been placed into the measuring cylinder with an uncertainty of $0.5cm^3$, the percentage error would be given by:

(0.5 / 25.0) x 100 = 2%

> **? Did You Know?**
>
> *Digital uncertainties*
> If you are curious why this is the case, the reason is that the balance or any other pieces of digital equipment takes a 'before' and 'after' reading. This will be the start reading, i.e., 0.000g and the final reading, 10.087g. Both of these readings have an uncertainty of half the smallest division, but as there are two readings they will add up to be the smallest division.

So we could quote the volume of water to be 25cm³ ± 2%.

As a one off value, this is not really very useful—it is far more useful to see the volume as 25.0cm³ ± 0.5cm³. However, the real advantage of the percentage errors is they allow you to determine an overall uncertainty value, when you have many 'non-connected' values.

This is best illustrated through our next example, Example 6.

> **Example 6:**
> A student is determining the enthalpy change for a reaction. This involves using a Propan-1-ol burner to heat up 50cm³ of water in a copper calorimeter. For reference, a photograph of this information is included opposite.

Using the electronic balance, 50.086g ± 0.001g of water is heated.

A temperature change of 18.65°C ± 0.05°C is recorded with an analogue thermometer.

We can calculate out the energy change for the reaction using q=mcΔT (c = specific heat capacity of water = 4.18 J mol^{-1} K^{-1}):

$$q = 50.086 \times 4.18 \times 19.65 = 4110 \text{ J (3 sf)}$$

Also see section 1.8 Significant Figures on the next page.

The percentage uncertainties for the balance and thermometer are:
Balance: (0.001 / 50.086) x 100 = 0.002%
Thermometer: (0.05 / 18.65) x 100 = 0.268%

These values can be added together, to give a total of 0.270%, so the calculated value is quoted as 4110 J ± 0.270%.

The percentage uncertainty can then be converted into an absolute uncertainty, in other words, 0.270% of 4110:

$$(0.270 / 100) \times 4110 = 11.0148 \text{ J} = 10 \text{ J (1 sf)}$$

It should noted that the accepted convention is to record the absolute uncertainty to 1sf and this gives a final value of:

$$4110 \text{ J} \pm 10 \text{ J}$$

The overall uncertainty allows us to get a feel for the overall calculated value and we know that, from our results, the value will be somewhere in the 4100 J–4120 J range. The significance of this becomes important later on in the conclusion section of the report when systematic and random errors are considered (see section 1.12 Systematic and Random Errors on page 19 for more information).

It should be stressed that this is a *simplified* example; more calculations would need to be done to work out ΔH in kJ mol^{-1}.

1.8 Significant Figures

As a rule, your final answer should be given to the minimum number of significant figures that have been used in the all of the calculations leading up to it.

Be careful not to round to early though—this can affect the final answer. Rounding should be the last thing you do.

Example 6 (page 17), is also good to illustrate the importance of significant figures in calculations. In this example we had:

$$q = 50.086 \times 4.18 \times 19.65 = 4110 \text{ J (3 sf)}$$

The minimum number of significant figures is three from the specific heat capacity, 4.18. Hence, the final answer is quoted to 3 sf.

Strictly speaking this should be written in standard form as:

$$411 \times 10^2 \text{ J}$$

This is because it is not clear if the zero in 4110 is significant or not. By putting it in standard form it becomes clear that the zero is not significant. Although strictly correct, this is an unusual way of presenting answers, but you can get around this problem by writing the number of significant figures at the end of the answer.

1.9 Conclusion

You should write a clear statement of your experimental results. Hopefully, this should link to what you were trying to establish in your research question.

Theoretical content is expected to arise in the conclusion, even if the assessment criteria do not make this obvious and you should relate what you have found out to the theory.

You may find that your results contain an anomaly—if this is the case, state it in the conclusion. You can discuss the reasons for the anomaly in the evaluation.

1.10 Literature Value

The literature value is an accepted value for what you are trying to find out. Ideally, your experiment will have a published literature value that is accessible.

The literature value can be found in a range of sources, depending on what you are trying to find out—the source of the literature value is not always obvious. The best place to start will be the IB data book but there are other data books available. You should always reference the source of your literature value.

Sometimes there may not be a literature value for your experiment. If this is the case, please state this. It is also important to understand the reliability of a literature value and you should always quote the source you have obtained the value from. Some examples are given below:

Experiment aim	Where the literature value can be found
Enthalpy of combustion	IB data book
Enthalpy of neutralisation	Chemistry Data book (e.g., Stark, Wallace, McGlashan)
Amount of vitamin C in orange juice	Side of container (nutritional information)
Order of a reaction	Text book / Research paper
Separation of amino acids by chromatography	Online data base: http://www.biotopics.co.uk /as/amino_acid_chromatography.html
Amount of Fe^{2+} in an iron tablet	Side of container

Sometimes there may not be a literature value for your experiment. If this is the case, please state this. It is also important to understand the reliability of a literature value and you should always quote the source you have obtained the value from.

1.11 Total Percentage Error

The Total Percentage Error compares your result to the data book result. It is a good starting point to get a feel for how reliable your experiment has been.

It is calculated:

$$\frac{\text{Experimental Value} - \text{Literature Value}}{\text{Literature Value}} \times 100$$

The value should always be positive, as it is a difference between the experimental and literature value, so quote it as such.

1.12 Systematic and Random Errors

1.12.1 Random errors

Random errors are errors that you have measured, recorded and taken into account—they are those that you have already calculated in your data processing, for example, the uncertainty value associated with a pipette.

1.12.2 Systematic errors

These errors are those that you have not measured or taken into account. They can be small, e.g. assuming a substance is 100% pure when it is actually 99% pure, or large, e.g., heat loss in enthalpy of combustion experiments.

Systematic errors are caused by a flaw or fault in the method or apparatus, in other words the 'system'.

1.12.3 Discussion of random and systematic errors

This is where all your hard work so far begins to come together.

By now, you should have calculated a value for your experimental work. Using our next example, this can be shown.

> **Example 7:**
> You have been determining the enthalpy of neutralisation of a strong acid and strong base and have calculated it to be -49.5 kJ mol^{-1} ± 9.2 kJ mol^{-1}.
> You have also been able to look this up in the data book and have found it to be -57.2 kJ mol^{-1}.
> You are now in a position to evaluate how accurate your experiment has been.
> According to your results, the calculated value could lie anywhere between -40.3 and -58.7 kJ mol^{-1}.

In this example you have obtained a pretty reliable result. The data book value lies within the error margins of your calculated value.

This means you have taken into account and measured most of the major sources of error. In other words, you have measured most of the random errors and they are larger than the systematic errors.

> **Key Point**
>
> If there was not a literature value for your experiment you will not be able to carry out this assessment. However, if you have a graph and a line of best fit you will be able to gauge / estimate the reliability of the experiment by looking at your data points and how close they lie to the line of best fit.

If, however, you did the same experiment but obtained a result of say, -23.8 kJ mol^{-1} +/- 5.8 kJ mol^{-1}, the data book value (-58.7 kJ mol^{-1}) lies outside of your range.

This will tell you that, although you have taken into account some experimental errors, there are other errors in your experiment that you have not taken into account.

In this case, the systematic errors will be larger than the random errors. If this happens do not be disappointed. Good scientific discoveries come about when something unexpected happens.

In your report you should comment on this and attempt to come up with some systematic errors that you have not measured or taken into account. In the example on the previous page, maybe there has been excessive heat loss or the acid was impure or the wrong concentration.

1.13 Evaluation (of the Method)

No good lab report is complete without a thorough evaluation of the method. The ultimate aim, which comes later in the Improvements section, will be to fine tune the experiment so that the data is as close as possible to the literature value.

You need to reflect on the method you used and try to identify possible systematic and random errors.

One of the hardest aspects in the evaluation is to get the 'direction' of error correct. This is best illustrated through Example 8:

> **Example 8:**
> You are preparing a sample of lead (II) iodide.
> To achieve this you have made a precipitate of lead (II) iodide by adding an excess of lead (II) nitrate to potassium iodide according to the following reaction:
> $$Pb(NO_3)_2(aq) + 2KI(aq) \rightarrow PbI_2(s) + K_2(NO_3)_2(aq)$$
> To obtain the sample you filter and gently dry the lead (II) iodide in an oven.

You have carried out your calculations and expected to obtain 1.52g of lead (II) iodide. In reality you obtain 1.28 ± 0.12g.

The result obtained is outside of the expected result so there are clearly systematic errors and your evaluation needs to identify some of these errors. Try to focus on what you think will be the major source of error.

In this example, it would be wrong to say that the sample was not fully dry (aside from the fact that you should dry to constant mass—but more on this in the Improvements section). If this were the case, it would be reasonable to expect the lead (II) iodide to have a greater mass than the expected 1.52g, not less—the direction of error has been misunderstood.

Continuing with this example, it would be correct to say that some lead (II) iodide was stuck to the filter paper and thus, when the mass of the precipitate was taken, the full mass was not measured as some lead (II) iodide was left behind (see the Improvements section for more information).

In this example, the direction of error is correct as the mass obtained would be less than the expected mass (and this is reflected in the data).

Both of these examples relate to systematic errors, but you should also include some random errors (this section of the evaluation crosses over into the improvements, see page 22 for more

> **Key Point**
>
> Try to be as critical as you can and allow for some thinking outside of the box. For example, did the lead (II) iodide decompose in the oven, losing mass due to iodine vapour being given off?

information). For example, an obvious improvement would be to repeat the experiment. However, this would make the results more accurate but not necessarily more reliable.

Another way of cutting down the systematic error would be to use a smaller measuring cylinder or a pipette.

For example, there will be more uncertainty in measuring 10cm³ of water in a 250cm³ measuring cylinder than 10cm³ of water in a 10cm³ measuring cylinder.

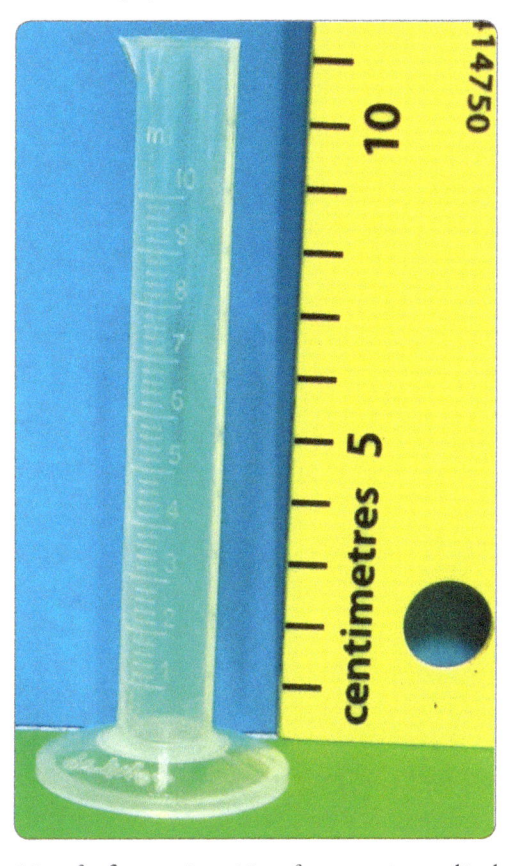

10cm³ of water in a 10cm³ measuring cylinder has an uncertainty of ± 0.1cm³ and so 10cm³ of water will have a percentage error of (0.1 / 10) x 100 = 1%.

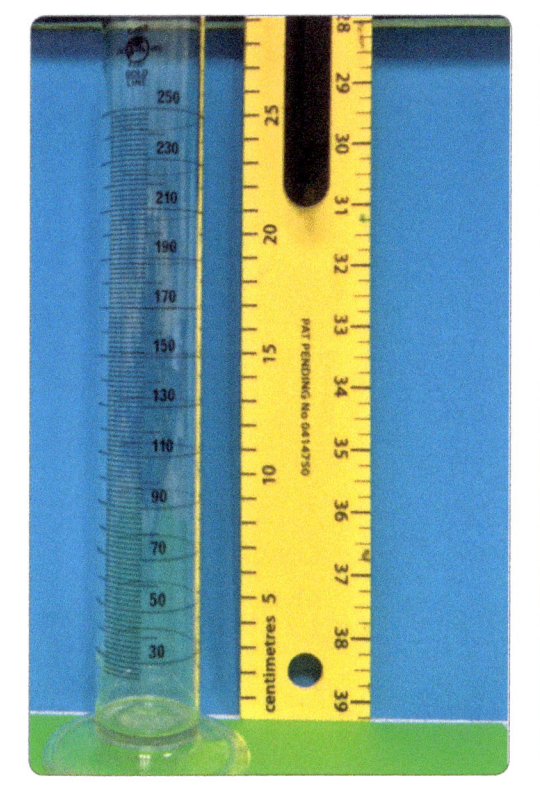

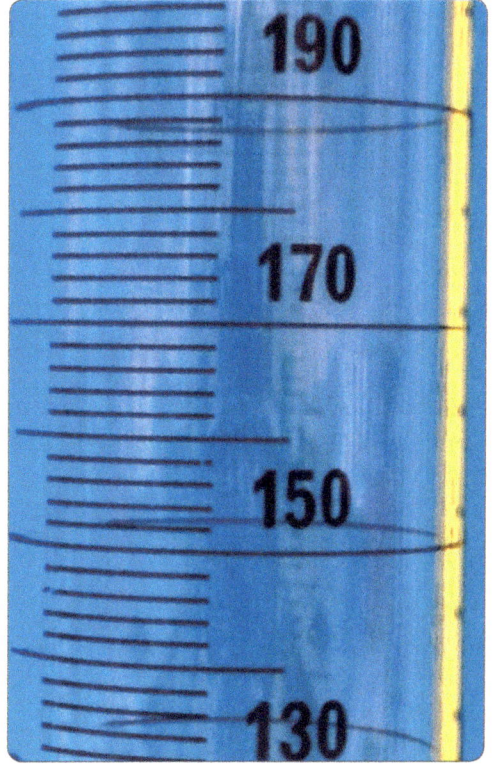

Compare this with 10cm^3 of water in a 250cm^3 measuring cylinder. This has an uncertainty of ± 1cm^3 and so 10cm^3 of water in this will have a percentage error of (1/10) x 100 = 10%.

So, just by using a better-sized or more appropriate piece of apparatus you can cut down on your random errors.

Your evaluation should be realistic. In Example 8 it would be unrealistic to state that the experiment was not finished and the reaction was still taking place, or that lead (I) iodide had formed. Even though the direction of error would be correct, it is not scientifically true as precipitation reactions are instantaneous and lead does not form 1$^+$ ions.

1.14 Improvements (to the Method)

As stated in the Evaluation section, the aim of the improvements to the method is to fine-tune things so that future experiments will achieve results closer to the literature value.

You now need to think of the method you used in your experiment—what aspects of this method were inefficient? What could you fine tune to improve the results? What potential improvements would these make to the original experiment?

Again, this is most easily illustrated though an example and we will use Example 8 (page 20), concerning the preparation of lead (II) iodide.

Two potential sources of effort were cited:
- Not heating the precipitate to constant mass, which would mean that the precipitate may still be wet when its mass is measured.
- The precipitate sticking to the filter paper, which would mean that the measured mass is less than it should be.

If the method did not state about heating to constant mass, it would be reasonable to add this to the improvements section, stating the reason why this is important (i.e., to ensure the sample is fully dry).

For the second point, rather than risk losing mass by transferring the lead (II) iodide from the filter paper to a weighing bottle, it would be much better to have pre-weighed the filter paper before filtering. The mass of filter paper would be known, so it would be a simple calculation to measure the mass of the precipitate by taking one value away from the other.

As with the evaluation, be realistic. Focus on aspects of the method you can control and improve. It would be unrealistic to say that the experiment could be improved by using purer salt solutions as you are unable to control this and your chemistry department is unlikely to have / be able to afford medicinal grade chemicals.

Here there is an opportunity to focus on cutting back the random errors (evaluation of the method on page 20 as well).

It is also possible to increase the quantities used while keeping the same apparatus; for example, using a burette with an uncertainty of ±0.05cm^3 to measure 50cm^3, rather than 5cm^3, will give a 0.1% error, rather than a 1% error.

1.14.1 A note on evaluations and improvements

It is very easy for the evaluation and improvements to blur into one if you are not careful. Keep them distinct. The evaluation is where you constructively criticise your method where the improvements state ways in which the faults in your method can be overcome.

> **Key Point**
>
> Make sure your evaluation and improvements are realistic and not superficial. Comments like 'I ran out of time' or 'use more accurate apparatus' are not going to impress your teacher or the moderator.

It is perfectly acceptable to put the evaluation and improvements in a table so that one will automatically feed into the other.

Using Example 8 (preparation of lead (II) iodide—page 20) this may look as follows:

Evaluation of the method	Improvements to the method
The method did not allow for the sample to be fully dried.	Heat the sample to constant mass.

1.15 Safety and the Environment

It is good practice to risk assess anything that you are carrying out as you bear responsibility to the safety of others, not to mention the safety of yourself.

If you are using a new piece of equipment seek advice. Ask your teacher how to use it and ask for a demonstration. Always wear safety goggles and a lab coat. Carry out lab work standing up and not sitting down.

When looking for hazards associated with chemical reagents, CLEAPPS 'hazcards' (http://www.cleapss.org.uk) are an excellent resource. Your school may be a member of CLEAPPS (or similar organisation) and may allow you access to the website or provide you with printed versions of the 'hazcards'. Make it a point of looking up chemical hazards—don't assume you know it as the hazcards are continually being updated.

As well as looking after each other, we also need to look after the environment. Just because you tip a chemical down the sink, it does not mean that it has gone forever. That reagent will make its way somewhere—probably ending up in a river or the sea and possibly being ingested by a plant or animal. Even though the reagent will have been diluted it can still cause problems, possibly killing an animal or plant, or entering the food chain and causing harm to other organisms in the food chain.

This means that it is your responsibility to ensure that reagents are correctly disposed of. Your school should have a system in place for the safe disposal of used chemicals—for example, potassium chromate (VI) should not go down the sink as it is a known carcinogen. Make it your goal to find out what your schools policy is and what it applies to.

Ultimately, it is your teachers' responsibility to ensure that any lab work you carry out is safe for you and your fellow classmates. This will include safe use of apparatus, as well as safe use of chemical reagents.

NOTES

2. THE ASSESSED CRITERIA

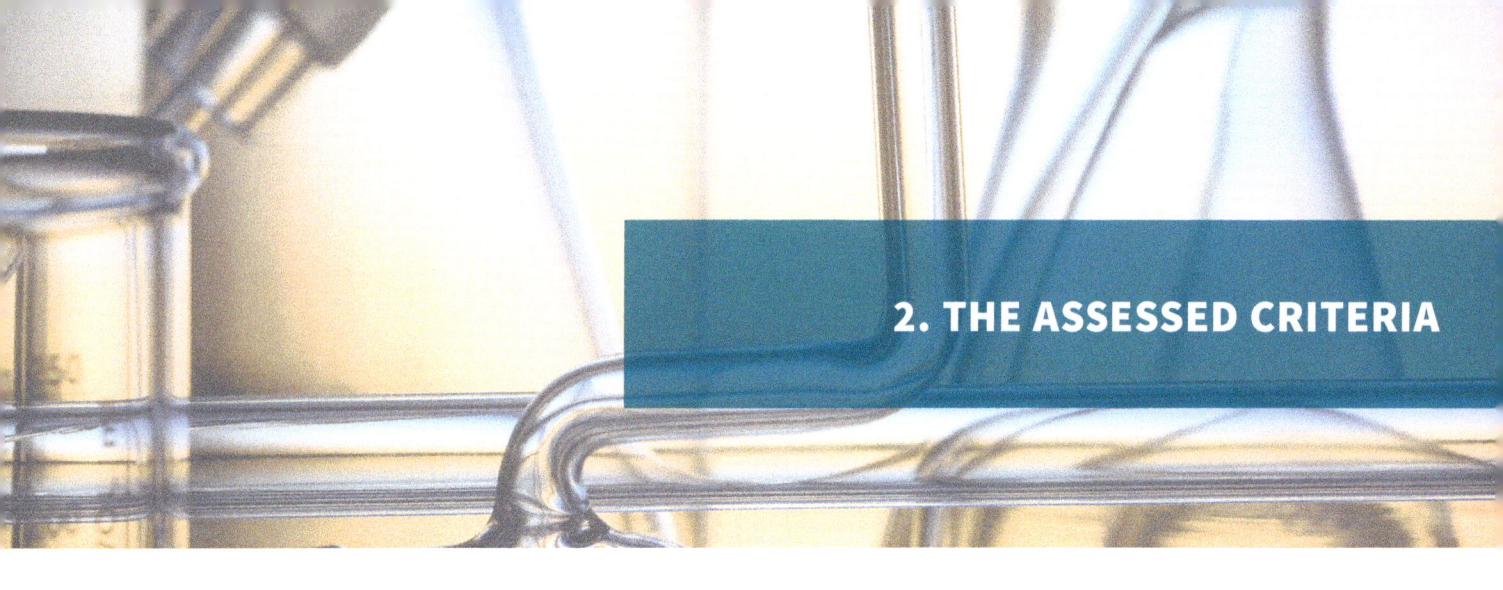

TOPICS:
Using the assessment criteria
Personal engagement
Exploration, Analysis
Evaluation, Communication

2.1 Introduction

Your Internal Assessment (IA) project counts towards 20% of your final grade. This is a sizeable proportion of your final grade and could be the difference between a '4' and a '6'. The IA consists of five different parts called 'criteria'. Each criteria has a maximum mark as follows:

Criteria	Maximum Marks
Personal Engagement (PE)	2
Exploration (EX)	6
Analysis (AN)	6
Evaluation (EV)	6
Communication (CM)	4
Total	24

2.2 How to Use the Assessment Criteria

You should ensure that you are familiar with the assessment criteria and understand how they are applied.

Each criteria has a number of statements that the IB refers to as 'descriptors', plus a mark band into which these descriptors fall.

To illustrate this we will use a simplistic, non-chemistry example. Let us imagine we are assessing a student's ability to play soccer (football). The criteria may look like this:

Mark	Descriptor
0	The student does not reach a standard given by the descriptors below.
1–2	Can pass the ball to another player some of the time. Can sometimes head the ball in a general direction. Can dribble the ball over a short distance. Understands a few of the major rules of the game.
3–4	Can pass the ball accurately to another player most of the time over short distances. Can head the ball powerfully or accurately. Can dribble the ball reasonably accurately over short distances. Understands most of the rules of the game and sometimes applies them in a match situation.
5–6	Can pass the ball to another player accurately over significant distances. Can head the ball powerfully and accurately. Can dribble the ball accurately over considerable distance. Understands all the rules of the game and can apply them in a match situation.

It is fairly clear to see that these criteria take into account four skills that a soccer player should have, for example, being able to pass, head, dribble and understand the rules.

If you have been asked to assess a student according to their ability to play soccer you will need to watch that student play and assess each of the attributes covered separately.

You may decide to highlight the descriptors as you assess the student to obtain something like has been done in the table below.

Figure 2.1: *Everybody loves football (soccer)* **Source:** *By Soccerfan1996 - Own work, https://commons.wikimedia.org/w/index.php?curid=50338968 [accessed 15 Feb 2017] (CC BY-SA 4.0).*

Mark	Descriptor
0	The student does not reach a standard given by the descriptors below.
1–2	Can pass the ball to another player some of the time. Can sometimes head the ball in a general direction. Can dribble the ball over a short distance. Understands a few of the major rules of the game.
3–4	Can pass the ball accurately to another player most of the time over short distances. Can head the ball powerfully or accurately. Can dribble the ball reasonably accurately over short distances. Understands most of the rules of the game and sometimes applies them in a match situation.
5–6	Can pass the ball to another player accurately over significant distances. Can head the ball powerfully and accurately. Can dribble the ball accurately over considerable distance. Understands all the rules of the game and can apply them in a match situation.

This soccer player can pass the ball and head it well, has a reasonable grasp of the rules but is not too good at dribbling the ball.

Ultimately you need to award the student a grade. The grade needs to be a whole number.

In order to do this, first of all, try to decide on a grade for each descriptor you have highlighted.

For example, the soccer player is really bad at dribbling, so this is a '1'.

They do show a sound grasp of the rules and you saw a few things that showed an understanding of how they are applied in a match situation. The result is the student gets a '4'.

The ball is passed accurately but not over huge distances, so a '5' is appropriate.

Finally, they head the ball very well so you decide on a '6'.

To award the final grade you use a best fit approach. The best fit awarded for the last three descriptors (4, 5, 6) would be a '5' but the '1' is going to pull this down so a '4' is probably the most appropriate grade to award.

This is how your teacher will decide on your grade—so to obtain a high mark you need to be hitting most of the high-end descriptors, not just some of them.

You may also find it useful to cross-reference these points with the information given in Chapter 2.

2.3 Personal Engagement (PE)

Summary: *This criterion assesses how well you have engaged with your own investigation.*

Mark	Descriptor
0	The report does not reach the standard below required for a '1'.
1	The investigation shows little evidence of the student coming up with ideas of his or her own for the practical work.
	There is no evidence that the research question has any relevance to the student.
	The exploration will be a 'standard' method and has not been adapted in any way by the student. The student has not been creative in planning, carrying out and writing up the lab work.
2	The investigation shows evidence that the student has come up with his/her own ideas for the practical work.
	The research question has relevance or significance to the student.
	The exploration is novel and the student has been creative in planning, carrying out and writing up the lab work.

Things to consider in your IA report:

Does your investigation demonstrate:

- Independent thinking?
- Initiative?
- Creativity?
- Personal input?

Does the choice of research question show personal significance / interest / curiosity with respect to the design of the method, choice of materials, interpretation of the data, conclusions made, and the evaluation of the method and data?

Syllabus Link

An explanation of IB descriptors and some bullet points to remember with each descriptor are listed below. The descriptors are not an exact copy of the descriptors in the Chemistry Guide (syllabus) but have been adapted to make interpretation easier for you.

Author's Tip

Your teacher will need evidence that your investigation demonstrates all of the key factors—ask yourself if you can see evidence of these factors in your written work.

2.4 Exploration (EX)

Summary: *This criterion assesses the quality of your research question, your method, and awareness of safety and the environment.*

Mark	Descriptor
0	The report does not reach the standard below required for a '1'.
1–2	The topic can be identified.
	The report contains an unfocused research question.
	Any background information supplied is not related to the investigation.
	The method only partially relates to the research question and few variables are addressed. Not enough variables are controlled and no repeats have been taken. Not enough tests have been carried out.
	The student has not taken into account major safety, ethical or environmental issues related to the method.
3–4	The topic can be identified.
	The report contains a research question that is only partially focused and lacks precision.
	Some of the background information is correct and helps the understanding of the IA.
	The method generally relates to the research question and some of the variables are controlled, or not all of the tests have been repeated, or not enough tests have been carried out.
	The report has taken into account most major safety, ethical or environmental issues related to the method.
5–6	The topic being investigated is clear.
	The report contains a fully focused research question that is prominent.
	The background information is correct and is related to the investigation.
	The method relates to the research question. All the variables are controlled and the correct number of trials have been carried out with repeats.
	The report has taken into account all the major safety, ethical or environmental issues related to the method.

Things to consider in your IA report:

- Is your research question relevant to the investigation?
- Is your research question fully focused?
- Is your research question clear?
- Have you included background information?
- And does it relate to the theoretical content?
- Is the method logical?
- Is the method easy to follow?
- Does the method list the independent, dependent and controlled variables?
- Have you carried out repeat measurements?
- Does the method take into account safety?
- Does the method take into account environmental issues?

Other things covered in the assessment of this criterion that are not obvious from the assessment criteria:

- Is the background information generalized or specific? It should be as specific as possible.
- Is the method too simplistic?
- Is the method imprecise?

- Are you carrying out a comparison of, for example, two brands? This is will be a weak research question as any differences will not be due to scientific principles.
- Have you included information about trial experiments?
- Have you included any information regarding calibration of apparatus?
- Have you referenced methods for the collection of data that you have taken from other sources?

2.5 Analysis (AN)

Summary: *This criterion deals with how well you record, process, interpret and conclude information from your collected data.*

Mark	Descriptor
0	The report does not reach the standard below required for a '1'.
1–2	There is not enough raw data to allow a conclusion to the research question to be written.
	A small amount of data processing has been carried out, but it is incorrect or there is not enough raw data to lead to a conclusion.
	Uncertainties or error analysis is minor or omitted altogether.
	Any data that has been processed has been interpreted wrongly, making the conclusion incorrect and lacking detail.
3–4	There is some quantitative or qualitative raw data that allows a basic conclusion to the research question to be written.
	Data has been processed and this leads to a generally valid conclusion. However, some of the data processing is not accurate or consistent.
	Processing of uncertainties or error analysis has been partially undertaken.
	Any data that has been processed has generally been correctly interpreted, allowing an incomplete conclusion to the research question to be made.
5–6	There is adequate quantitative or qualitative raw data that allows a full conclusion to the research question to be written.
	The correct amount of data has been processed accurately and this leads to a valid conclusion being made that is in line with data from the lab work.
	Processing of uncertainties or error analysis has been fully and correctly undertaken.
	All data that has been processed has been correctly interpreted, allowing a full conclusion to the research question to be made.

Things to consider in your IA report:
- Have you included quantitative raw data?
- Have you included units?
- Have you included qualitative raw data?
- Have you recorded uncertainties?
- Is your data recorded to the correct precision?
- Have you correctly processed your raw data?
- Have you included a worked example of how you have processed data?

Other things covered in the assessment of this criterion that are not obvious from the assessment criteria:
- Have you drawn a good graph, with and x and y axes, as well as major and minor gridlines?
- Have you drawn a line / curve of best fit?

- Have you identified any outliers on your graph (if appropriate)?
- Avoid pages and pages of raw data—a sample will suffice.
- Can you see and identify patterns in your data?

2.6 Evaluation (EV)

Summary: *This criterion deals with your evaluation of the method and suggested improvements.*

Mark	Descriptor
0	The report does not reach the standard below required for a '1'.
1–2	The stated conclusion does not relate to the research question or the data collected and / or processed.
	There is little or no comparison to the scientific context (literature value) of the investigation.
	An evaluation of the entire investigation is outlined and only relates to the method.
	Any improvements or extension to the investigation stated are brief.
3–4	The stated conclusion relates to the research question and this is reflected in the data collected and / or processed.
	The conclusion relates to the scientific context (literature value) of the investigation.
	An evaluation of the entire investigation is thorough and relates to the method and restrictions associated with the data.
	Some stated improvements and extension to the investigation are thorough and achievable
5–6	The stated conclusion is detailed and completely relates to the research question. This is fully reflected in the data collected and / or processed.
	The conclusion is thorough and relates to the scientific context (literature value) of the investigation.
	An evaluation of the entire investigation is thoroughly discussed. There is a full evaluation of the method and restrictions associated with the data.
	Any stated improvements and extension to the investigation are thorough, achievable and relevant.

Things to consider in your IA report:

- Have you written a conclusion?
- Is your conclusion justified (does it relate to the data)?
- Have you considered the strengths of the investigation?
- Have you considered the weaknesses of the investigation?
- Are the weaknesses consistent with the direction of error?
- Have you considered improvements to the method?
- Have you considered extending the investigation?

Other things covered in the assessment of this criterion that are not obvious from the assessment criteria:

- Have you included a statement as to whether your data answers, or does not answer, the research question?
- Does the data support the hypothesis, if you have proposed one?
- Is the scientific context of the conclusion correct?

2.7 Communication (CM)

Summary: *This criterion deals with the effective presentation and communication of the investigation.*

Mark	Descriptor
0	The report does not reach the standard below required for a '1'.
1–2	The report is not very well presented which makes it difficult to interpret its scope and finding. The layout of the report is not clear. Information on the scope and its findings seems out of order or missing sections. There is unnecessary information in the report that makes it hard to find the scope and findings. Subject specific terminology and scientific or mathematical conventions are incorrectly used.
3–4	The report is not well presented, making it easy to interpret its scope and findings. The scope and findings can still be understood if errors have been made. The layout of the report is clear and the report flows. Information on the scope and its findings are presented in a logical and easy to understand form. There is no or little unnecessary information in the report that makes it straightforward to find its scope and findings. Subject specific terminology and scientific or mathematical conventions are correctly used or used with few or minor mistakes allowing the report to be easily understood.

> **Author's Tip**
> Write your method like a recipe—in other words, a step-by-step list of instructions. Number the steps. Again, it ensures clear and concise communication.

Other things covered in the assessment of this criterion that are not obvious from the assessment criteria:

- Passive voice v Personal voice—the IB does not mind which style you employ but be consistent. If you use the passive voice, ensure that it is passive throughout the report. Similarly, if you use the personal voice, make sure it is personal throughout.
- Length: the report should be between 6–12 pages in length, written in font size 10–12.
- Graphs need to be readable and not miniaturised so they cannot be read.
- Full calculations are not expected, although it would be wise to show a sample calculation.
- Appendices that take the page count over 12 pages should not be included as these will not be read or credited.
- Excessive quantities of raw data from a data logger should also not be included.
- Pay close attention to ensuring that the correct use of formulae, nomenclature, mechanisms, uncertainties and significant figures.
- Chemical equations should be balanced.
- As much as possible, use SI units throughout the report.
- Ensure you source / reference as fully and as consistently as possible all work that is not your own.

NOTES

3. WHAT MAKES A GOOD IA?

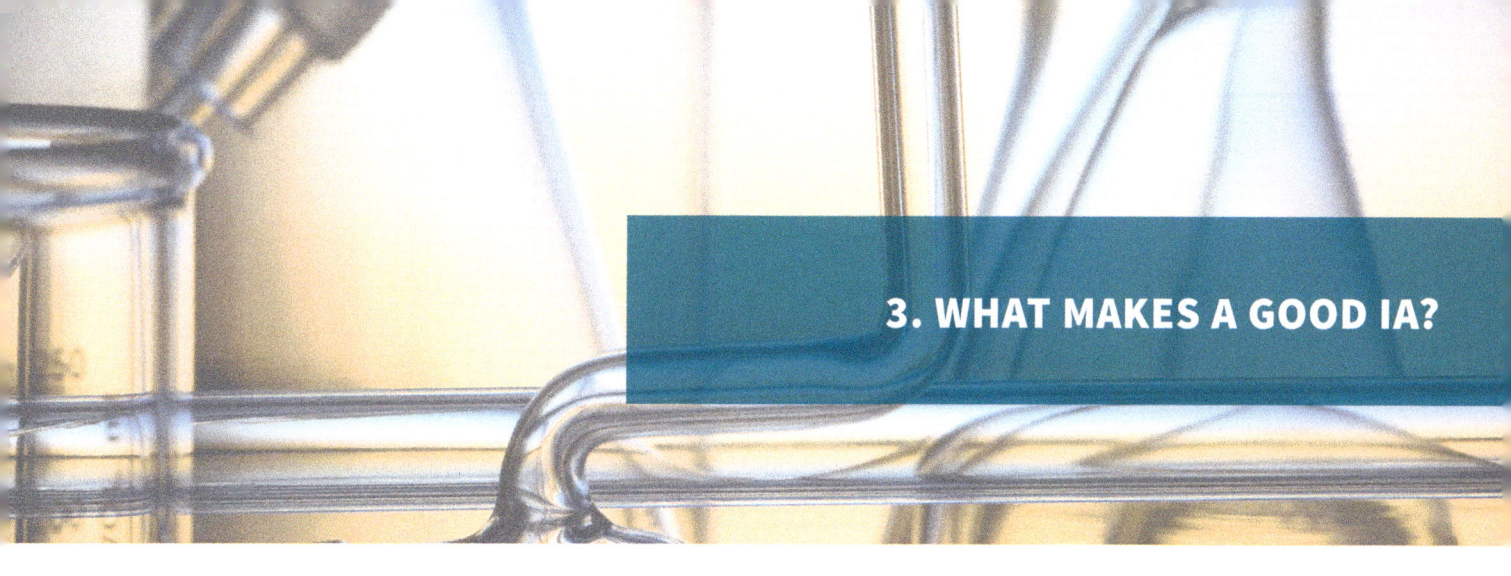

TOPICS:
Different types of IA
Ideas for IAs
Dos and don'ts for an IA

The key to producing a good IA is to come up with a *good research question*—this can't be stressed enough.

If your research question is sharp and well defined, then, due to the way that the assessment criteria are written, you will have a better chance of getting a good grade compared to a student with a poor research question.

This, however, is only half the battle.

The other half involves coming up with a topic that is going to allow you to generate a good amount of data through your focused research question.

Ideally, your IA will involve testing a variable a number of times, repeating these tests and then processing the raw data.

And here lies the challenge. It is very likely that during the teaching of the course that either you will be taught a concept or carry out a practical that really grabs your attention. That light bulb comes on and you actually think to yourself, 'wow, that's quite interesting!'

These 'wow' moments are the ones to hold onto—before you forget what it is, write it down, have a section in the front of your file that has the title 'IA ideas' and scribble it there.

But don't stop with one 'wow' moment; aim to write as a many as you can. This means that when the time for your IA comes:

1. You will have some ideas and you will not be starting from scratch.
2. You have a plan 'b' and 'c' and 'd' and so on—there may be a problem with one of your 'wow' moments. For example, maybe the school does not have the apparatus required, maybe the experiment takes too long, or is too dangerous. There could be a whole range of reasons why your idea can't be a reality.

When it comes to planning the IA, keep it simple. The best IAs are the elegant ones, those that take an existing idea and put a certain spin on it. For example, if you were looking at the effect of the concentration of a salt solution on the amount of metal deposited through electroplating, you could also look at the absorbance of the salt solution using a spectrometer to see if there was a relationship between the amount of metal deposited and the absorbance of the solution. Try to be creative.

You will not gain extra marks using costly or high-tech lab equipment and, similarly, you will not be penalised for using simple or more straightforward lab equipment. Indeed, you may find that your PE mark is higher as it involves more creativity from yourself.

3.1 Different Types of IA

It is permissible to carry out IAs that involve an aspect of theoretical chemistry—the IA does not necessarily need to be 'wet' chemistry.

There are five different types of IA allowed by the IB:
- Traditional / Hands on (e.g., aspirin purification by melting point)
- Spreadsheets (e.g., pI of proteins)
- Databases (e.g., investigating intermolecular forces)
- Simulations (e.g., using a virtual lab to determine the equivalence point in the titration of 10 mol dm^{-3} NaOH and 1.0 mol dm^{-3} CH$_3$COOH)
- Computational modeling (e.g., comparing predicted bond angles with true bond angles).

It should also be noted that your IA does not need to be on just one of these categories, for example, it could be traditional, 'hands on lab' that uses a spreadsheet for data processing and a database to compare calculated values with theoretical values.

3.2 50 Ideas for IAs

Below are some ideas for IA investigations that may be a useful starting point when it comes to planning / deciding upon a research question.

It should be noted that these are ideas and not research questions. The statements are deliberately broad and, if you decide to use any of these ideas, you should be a specific as you can.

Please make sure you show your idea to your teacher before you get too involved in the planning stages as there may be a very good reason why you are unable to carry this out.

1. Investigate chemical equilibria (determine Kc for a reaction).
2. Investigate a weak acid-strong base titration (determine the Ka of a weak acid).
3. Investigate the enthalpy change of a redox reaction.
4. Investigate an aspect of a simple voltaic cell.
5. Investigate the Ideal Gas Laws using a data logger.
6. Determine the amount of copper in coins using colorimetric analysis.
7. Analysis of seaweed (seaweed is a good source of Br$^-$ and I$^-$).
8. Investigate the amount of CaCO$_3$ in brown and white eggshells.
9. Compare the percentage of vitamin C in various brands of juice.
10. Investigate hardness of water from different local sources.
11. Investigate the speed of neutralisation of antacids from different sources.
12. Investigate the kinetics of the bromine clock reaction to determine the order of reaction.
13. Investigate factors affecting electroplating.
14. Investigate factors affecting electrolysis.
15. Investigate the % of chlorine in different bleaches or swimming pool water on different days.

> **Author's Tip**
>
> You should aim to keep this running in the back of your mind at all times and not just in school—who knows where one of those 'wow' moments will creep up on you (you could be brushing your teeth, watch TV, playing sport, who knows) but when it happens—write that thought down as soon as it is possible to do so.

16. Investigate the enthalpy of neutralisation of different acid / base concentrations.
17. Determine the activation energy of iodine-clock reaction.
18. Investigate the effect of pH on the rate of rusting.
19. Calculate the K_w for water at different temperatures by measuring its pH.
20. Investigate factors determining heat of combustion in alcohols.
21. Investigate the level of unsaturation in different oil brands.
22. Investigate the pH of soil.
23. Prepare and test of buffer solutions.
24. Investigate the effects of heterogeneous catalysis on the activation energy of a reaction.
25. Investigate how the hardness of water affects the solubility of salts.
26. Investigate whether oxidation of tea / coffee changes its pH.
27. Investigate the effect of soil pH on the chlorophyll content in plant leaves.
28. Determine the enthalpy change of the thermal decomposition of sodium hydrogen carbonate.
29. Determine the percentage of iron in an iron compound by a redox titration.
30. Determine the solubility of an ionic salt in water by plotting a solubility curve and compare the solubility with its literature value.
31. Investigate the effect of the length a salt bridge has on the dipped voltage produced by a voltaic cell.
32. Investigate a factor influencing retention in paper chromatography of amino acids.
33. Determine the calcium content of milk by EDTA titrations.
34. Determine the concentration of biodiesel produced from fruits by using calorimetry.
35. Investigate the energy densities of fuels versus the bond strength (using a database).
36. Measure the concentration of zinc ions in a dietary supplement.
37. Investigate variables affecting water absorption polymers.
38. Investigate the rate of evaporation of a liquid with respect to its temperature.
39. Investigate the effect of different sacrificial metals on the rusting of iron.
40. Investigate the rate of adsorption of organic acids on charcoal.
41. Investigate the factors affecting the colours of transition metal compounds (through the use of a colorimter).
42. Investigate the factors affecting the rate of reaction between iodine and propanone.
43. Investigate an equilibrium position and Le Chatelier's principle spectroscopically.
44. Investigate the catalytic abilities of different transition metal oxides.
45. Investigate the effect of solvents in paper chromatography and thin layer chromatography.
46. Prepare esters.
47. Investigate the solubility of salts (K_{sp} values).
48. Investigate the equivalence point of weak / strong acids and weak / strong bases.
49. Investigate the esterification of soaps.
50. Investigate the effect of different ionic salts on the freezing point depression of water.

Disclaimer:

You should always get your IA checked out by your class teacher before you carry out any lab work to ensure that your experiment is safe.

Any IA lab should be thoroughly risk assessed by yourself and your teacher should check this risk assessment.

3.3 Ten Dos and Don'ts for Writing a Successful IA

DO

- Plan ahead.
- Keep thinking about your IA—from the moment you first read this Guide.
- Read science articles for inspiration. Some examples include the BBC science section, the Royal Society of Chemistry or New Scientist magazine.
- Keep it simple.
- Make it yours. Make it original.
- Be clear on what you are trying to do—this will help form the focused research question.
- Source / reference any method you have found. Include a brief, concise and clear bibliography at the end.
- Speak to your teacher about what you are thinking of doing.
- Check the safety precautions / environmental hazards of any reagents / chemicals you use—use your school's hazcards.
- Ask for help if you are using an unfamiliar technique or unfamiliar apparatus.

DON'T

- Think a title is going to come to you on the day you start the lab work—you will need to think about it and plan ahead. Nobody said it would be easy.
- Use the same IA as your EE—a big no.
- Repeat a practical you have already carried out in school.
- Share results with someone else.
- Make results up or invent your own results.
- Be vague in your research, practical work or write up—be precise.
- Pick a practical with a result that is already widely known (for example, determining the enthalpy of combustion of ethanol—this can be sourced in the IB data book).
- Overcomplicate things.
- Plan to use reagents, chemicals or apparatus that are too expensive or are too dangerous (for example, an NMR spectrometer, cesium or fluorine).
- Think you can write up the investigation in one go and submit it—be prepared to draft, redraft and redraft until you are happy with the quality of it. Your teacher is permitted to read over one draft only and to make some general (unspecific) comments.

Appendix 1: List of Mandatory Practical Work as Specified by the IB

Topic	Descriptor
Topic 1.2	The obtaining and use of experimental data for deriving empirical formulas from reactions involving mass changes.
Topic 1.3	Use of the experimental method of titration to calculate the concentration of a solution by reference to a standard solution
Topic 1.3	Obtaining and the use of experimental values to calculate the molar mass of a gas from the ideal gas equation.
Topic 5.1	A calorimetry experiment for an enthalpy of reaction should be covered and the results evaluated.
Topic 6.1	Investigation of rates of reaction experimentally and evaluation of the results.
Topic 8.2	Candidates should have experience of acid-base titrations with different indicators.
Topic 8.3	Students should be familiar with the use of a pH meter and universal indicator.
Topic 9.2	Performance of laboratory experiments involving a typical voltaic cell using two metal/metal-ion half-cells.
Topic 10.1	Construction of 3D models (real or virtual) of organic molecules.
Topic 15.1	Perform lab experiments which could include single replacement reactions in aqueous solutions. (either here or 19.1)
Topic 19.1	Perform lab experiments which could include single replacement reactions in aqueous solutions.

Appendix 2: Table of Uncertainty Values for Lab Equipment Without a Scale

Apparatus	Size/Precision	Uncertainty
pipettes (grade B)	10 cm^3	± 0.02 cm^3
	25 cm^3	± 0.05 cm^3
volumetric flasks (grade B)	250 cm^3	± 0.2 cm^3
	100 cm^3	± 0.1 cm^3

Source: M. Bluemel, IB Chemistry: Student Guide for Internal Assessment (Oxford: OSC, 2010), p. 7.

NOTES

NOTES

NOTES

NOTES

NOTES

NOTES

NOTES

NOTES

IBDP REVISION COURSES

Summary

Who are they for?
Students about to take their final IBDP exams (May or November)

Locations include:
Oxford, UK
Rome, Italy
Brussels, Belgium
Dubai, UAE
Adelaide, Sydney & Melbourne, AUS
Munich, Germany

Duration
2.5 days per subject
Students can take multiple subjects

The most successful IB revision courses worldwide

Highly-experienced IB teachers and examiners

Every class is tailored to the needs of that particular group of students

Features

- Classes grouped by grade (UK)
- Exam skills and techniques – typical traps identified
- Exam practice
- Pre-course online questionnaire to identify problem areas
- Small groups of 8–10 students
- 24-hour pastoral care.

Revising for the final IB exams without expert guidance is tough. Students attending OSC Revision Courses get more work done in a shorter time than they could possibly have imagined.

With a different teacher, who is confident in their subject and uses their experience and expertise to explain new approaches and exam techniques, students rapidly improve their understanding. OSC's teaching team consists of examiners and teachers with years of experience – they have the knowledge and skills students need to get top grades.

The size of our Oxford course gives some particular advantages to students. With over 1,000 students and 300 classes, we can group students by grade – enabling them to go at a pace that suits them.

Students work hard, make friends and leave OSC feeling invigorated and confident about their final exams.

We understand the needs of IBDP students – our decades of experience, hand-picked teachers and intense atmosphere can improve your grades.

" I got 40 points overall, two points up from my prediction of 38, and up 7 points from what I had been scoring in my mocks over the years, before coming to OSC. Thank you so much for all your help! "

OSC Student

Please note that locations and course features are subject to change - please check our website for up-to-date details.

Find out more: osc-ib.com/revision +44 (0)1865 512802

MID IBDP SUMMER PROGRAMMES

Summary

Who is it for?
For students entering their final year of the IB Diploma Programme

Locations include:
Harvard and MIT, USA
Cambridge, UK

Duration
Min. 1 week, max. 6 weeks
1 or 2 IB subjects per week

- Improve confidence and grades
- Highly-experienced IB teachers and examiners
- Tailored classes to meet students' needs
- Wide range of available subjects
- Safe accommodation and 24-hour pastoral care

Features

- Morning teaching in chosen IB subject
- 2nd IB subject afternoon classes
- IB Skills afternoon classes
- One-to-one Extended Essay Advice, Private Tuition and University Guidance options
- Small classes
- Daily homework
- Unique IB university fair
- Class reports for parents
- Full social programme.

By the end of their first year, students understand the stimulating and challenging nature of the IB Diploma.

They also know that the second year is crucial in securing the required grades to get into their dream college or university.

This course helps students to avoid a 'summer dip' by using their time effectively. With highly-experienced IB teachers, we consolidate a student's year one learning, close knowledge gaps, and introduce some year two material.

In a relaxed environment, students develop academically through practice revision and review. They are taught new skills, techniques, and perspectives – giving a real boost to their grades. This gives students an enormous amount of confidence and drive for their second year.

> "The whole experience was incredible. The university setting was inspiring, the friends I made, and the teaching was first-class. I feel so much more confident in myself and in my subject."
>
> OSC Student

Please note that locations and course features are subject to change - please check our website for up-to-date details.

Find out more: osc-ib.com/mid +44 (0)1865 512802